Bird from HELL and Other Mega Fauna

Bird from HELL and Other Mega Fauna

SECOND EDITION

GERALD MCISAAC

Printed in the United States of America

ISBN 978-1-957009-83-4 (sc)
ISBN 978-1-957009-84-1 (hc)
ISBN 978-1-957009-85-8 (e)

Library of Congress Control Number: 2022923851

History
2023.01.09

CONTENTS

INTRODUCTION

Allow me to start by saying that I am a Caucasian male with a scientific background. About forty years ago I got tired of big city life and opted for life in the mountains. It is not a decision I regret. On the contrary, it is a life style I love.

I have since married an Indigenous girl, a member of the Sekani tribe, of the Dene people. We have raised a family on a Reserve referred to as Tsay Keh Dene, or Mountain People, loosely translated. The village is located in British Columbia, in the Rocky Mountain Trench, 350 kilometres or 200 miles from the nearest town, that of MacKenzie. The closest neighbouring Reserve is that of Kwadacha, 75 kilometres or 50 miles away.

Out of respect for our American neighbours, most of whom are not familiar with metric measurements or of Celcius, I have chosen to provide distances in imperial as well as metric, and temperatures in Fahrenheit as well as Celcius. Further, that which is referred to as a Reserve in Canada, is referred to as a Reservation in America. As most people who live in the United States refer to themselves as Americans, I have chosen to use that terminology, at the same time referring to the country as America.

The social structure on the Reserve is different from that to which most people are accustomed. Without going into any great detail, suffice it to say that such a social structure gives rise to very strong social bonds. Also, the names of the deceased are not to be mentioned, so I do not refer to them directly.

The older members of the Band, as the community refers to itself, are referred to as "Elders", and are regarded with the utmost respect. They are the equivalent to those referred to as "Seniors" in conventional society.

The Elders are the absolute experts on the mountains, and I use the word "mountains", in the loosest possible sense. This is to say that I include the forests, meadows, swamps, lakes and rivers. The Elders have provided me with detailed descriptions of animal which I recognize. Yet, it is the scientific opinion that those animals are extinct.

I maintain that the Indigenous Elders are correct, and the scientific community is completely mistaken. Now it is the—not so little matter—of proving this.

There is a sense of urgency in this, and not only because we are entitled to our wild life. It is part of our heritage. Also because many of these animals are predators. All predators have a keen sense of smell, and these predators are no exception. Further, all predators are far more likely to attack when they smell blood. That is the reason they prey mainly upon girls of child bearing age. It is absolutely essential that all members of the public be made aware of their existence.

It is correct to regard this as a book of science, which it is, one which is written in a popular manner, complete with some rather poor jokes. This stands in contrast to most books of science, which few common people read, if only because they cannot understand them. Yet as the scientists are not doing their job, including that of documenting the existence of these animals, then it is up to common people to take up the slack. Of necessity, certain technical terms are mentioned, and properly explained.

I can only hope that a great many people will be motivated to assist in locating these magnificent animals. You will not be disappointed.

CHAPTER 1

Background

The Dene Elders grew up in an environment which was very similar to that which is commonly referred to as the Stone Age. In fact, it was a time of great change, of cultural conflict, as it is now. They were being introduced to civilization, complete with modern tools and weapons, as well as diseases. They were also expected to adapt, to change their social and cultural way of life, in order to fit into civilized society. That same civilized society had no intention of adapting to the life style of Indigenous People. In fact, neither culture made any great effort to understand the other.

The Dene Elder were members of a hunting—gathering society. It is an understatement to say that life was difficult. All food had to first be gathered or killed. All tools had to first be fashioned. They lived constantly on the edge of starvation. That changed dramatically, at the time when the Dene first met traders from civilization. Those traders offered burlap sacks of flour, rice and sugar, in exchange for peltries, otherwise known as furs. This is to say that the traders were interested in the hides of animals, such as marten, mink, beaver, lynx, wolf, wolverine, coyote, grizzly and such.

The Dene had their own priorities. They knew how to weave baskets, and used these baskets, especially for gathering berries. These baskets were quite useful, but rather feeble. They were not capable of carrying anything much heavier than berries. By contrast, the rice, flour and sugar came in containers of burlap bags. As burlap is very strong, the Dene traded peltries for burlap bags. They had no interest in the contents, if only because they did not recognize it as food. The flour and rice were thrown on the ground, while the sugar was thrown into the fire. They liked to see it sparkle as it burned. Entertainment!

Further, the Dene were at such a primitive stage of development, that they had no pottery. For that reason, the mere act of boiling meat was difficult. The only way in which this could be done, was by using the stomach of an animal, as a container. The water was added to the container, hot rocks from the campfire were tossed into the water, and the meat was added. Mountain goat was considered the best for this, as they were able to get three meals, from each container, in this manner.

This should give some idea of the world in which the Dene lived. The resultant cultural confusion is understandable, and persists to this day. Of course they were nomads, following the herds, unable to stay in one place for any length of time, if only because their food supply would soon be exhausted. Their survival depended largely upon their knowledge of the animals, and their habits. The fact that they survived, is proof that they are the experts, in regards to these mountains. For that reason, I pay strict attention to that which they tell me.

Numerous people have questioned the sanity of the Dene. They point out that those who choose to live in such a harsh environment, must be crazy. To such people, I can only respond that there was not a great deal of choice in the matter.

It is the scientific opinion that there were possibly three migrations of people across the Bering Strait, from Asia, over a period of possibly thousands of years. The Rocky Mountain Trench provides a natural corridor. The ancestors of the people with whom I live, apparently came across the Strait in the most recent migration, possibly around the time of Columbus, within the last few hundred years.

These "immigrants" soon found themselves in a situation which is similar to that which modern day immigrants face. In other words, "not welcome"!

The Dene immediately found that the finest "real estate", such as the coastal regions, was already occupied. Such regions provided the inhabitants with a great abundance of seafood, including fish, whale, walrus, seal, sea lion and dolphin. Also fresh water fish, as well as deer, moose, elk, bear, duck, goose and so forth. In addition, the surrounding forests of cedar and hemlock provided the material for lodging and tools. Such coastal areas were as highly prized then, as they are now. Further, the people who lived in those areas were not at all anxious to "share their wealth".

That left the land to the east, the prairies, with vast herds of bison and other wild life, but that too was occupied. Those people too, had no intention of sharing. The Dene, as the latest immigrants, had no choice but to subsist, as best they could, in the corridor, the Rocky Mountain Trench.

The Trench is famous for its magnificent scenery and big game. Professional photographers have taken countless pictures of the mountains, while trophy hunters are enthusiastic about killing the biggest male animals, such as moose, elk, caribou, mountain sheep, mountain goats and bears. Grizzly and mountain sheep are the most prized trophy animals, and the hunters pay a small fortune, for a chance to kill one of those animals.

The Dene who live in the mountains are more impressed by the difficulty in surviving. The mountains are truly magnificent, as the poets and photographers can testify. It is also a fact that those same mountains are not the slightest bit forgiving. Any moment of complacency can prove to be fatal. Recently, a couple of girls were reminded of this.

They decided to head for town in the middle of winter, with the temperature close to forty degrees below zero. At that temperature, Fahrenheit and Celsius are almost identical, supremely cold. Their vehicle broke down and they had no two-way radio, so no way to call for help. That and the fact that they had no matches, no way to light a fire, meant that they were in serious trouble. They spent a severely cold night in the pickup and were found the next day. They could not have survived another night in that severe cold.

The only reason they were not rescued immediately, was because no one was looking for them. The first vehicle that came along stopped, and offered assistance. In the mountains, this is common practice. We tend to take care of each other. After all, the roads in these mountains are not to be confused with highways.

Those two girls lived in these mountains all their lives, and had driven those roads many times, without any major mishap. As a result of this, they developed a certain complacency, which nearly cost them their lives. Of course they should have carried with them a box of matches and candles, as well as a two-way radio. Under such circumstances, the vehicle provides the shelter and the lit candles provide sufficient heat, to keep body and soul united.

It was only within the last thirty years, that roads and bridges have been built, connecting the village with the highway. Electricity is supplied by generators, and the village now has running water, a modern school, clinic and store. This has resulted in a vast improvement in the living standards of the people who live there.

The roads leading into the village are maintained by the Forest Service. These are commonly referred to as "dirt roads", while in fact they are gravel roads. They are also sometimes referred to as logging roads, although gravel trucks, mining ore trucks and low bed trucks also use them. Such vehicles are referred to as "trucks", while passenger vehicles are referred to as "pickups". Very few cars travel these roads, as they cannot take the punishment.

These roads tend to be rather narrow, with the occasional "wide spot" that are suitable for pulling into, when approached by a "wide load". A loaded logging truck or ore truck, as well as a low bed truck, carrying a big machine, qualifies as a "wide load".

At the start of each road, signs are posted. These signs state the name of the road, as well as the correct radio frequency to be used.

As well, kilometre signs are posted every kilometre or two. Those who travel these roads are expected to have a two-way radio, to announce their location and direction of travel, on a regular basis, perhaps every couple of kilometres. Vehicles which are "loaded", which generally means going in the direction of town, have the right of way. It is up to vehicles which are "empty", or going in the opposite direction, to get out of their way.

Until quite recently, the various "Main Lines", or logging roads, were assigned certain radio frequencies, and these frequencies were assigned various names. As these names and frequencies varied across the province, it led to a certain amount of confusion. When driving these roads, the last thing anyone needs is confusion. That can cost someone their life. For that reason, the Forest Service decided to change the system, making it standard. Now each radio frequency is given a number, and these numbers are used throughout the province. This simplifies the matter, but also meant that everyone had to buy a new radio. It is a small price to pay for safety. As well, the words loaded and empty have been replaced with "up" and "down".

As an example, a logging truck travelling south to the mill, with a load of logs, on the Finlay Main Line, may see a sign that reads "18km". All drivers know that km is short for kilometres, so the driver is expected to get on the two-way radio and announce "eighteen down on the Finlay". If the driver is in a pickup, then that should also be specified. Any vehicles going in the opposite direction, "up vehicles", are expected to pull over into a wide spot, to allow the "down vehicle" to pass. It should be noted that people who use these roads, tend to refer to kilometres as "clicks".

As for those who consider this to be somewhat unfair, that a loaded truck has the right of way, consider the fact that each logging truck may carry fifty tons of logs. If the truck makes a sudden stop, then the logs it is carrying tend to come forward into the cab of the truck. Such things do happen, and they tend to ruin the whole day of the driver.

The side roads, which branch off the various main lines, lead to logging areas, which we refer to as "logging blocks". After the timber is harvested, the roads inside that block are "deactivated". This means that a machine, usually a backhoe, goes into the area and tears up the road at regular intervals. As well, the bridges are usually removed. This generally happens only after the block has been replanted.

The road between the two Reserves of Tsay Keh Dene and Kwadacha, is called the Russel Main Line. It follows the Finlay River, but on the west side of the river. On the east side of the river, there is another Main Line, called the Finlay Main Line. Traffic on the Russel Main Line uses a different radio frequency than traffic on the Finlay Main Line. In each case, the vehicles are expected to specify which road they are travelling. This is in the interests of safety, as we all make mistakes. "Down" vehicles are especially expected to announce their location, while "up" vehicles may also announce, if only less frequently.

No doubt, there are a great many people who have a difficult time imagining life in this remote area. Those who first move up here, especially teachers, tell me that the culture shock is severe. It is similar to moving half way around the world! The point being that if the reader has a difficult time in imagining life in these mountains, that is completely understandable. For that reason, I am going into such detail. These mountains are rugged and remote, and are home to several species of animals that are thought to be extinct. They are rarely seen, if only because there are so few people living in these mountains.

Many years ago, I became convinced of the existence of these animals, and decided to devote my life to proving this. My method involves separating fact from belief. I focus on the description of the various animals, which tends to be detailed and accurate. At the same time, I set aside the personal beliefs of the individual, who is describing the animal. I respect the beliefs of all common people, members of the public, while not necessarily sharing those beliefs. I have my own beliefs. In turn, I expect all others to respect my beliefs.

This brings me to the scientists. They are certainly not "common people", "members of the public". On the contrary, they are highly trained professionals. They too have presented various theories, which they regard as facts, not to be challenged. True, bones and fossilized remains are facts. They speak for themselves. From these facts, the scientists have drawn certain conclusions, which they believe to be correct. These conclusions, or beliefs, are referred to as theories. These theories are just that, *beliefs, which are presented as facts!*

Scientific theories are meant to be challenged! That is the one and only way, in which we learn about the world in which we live! These theories should not be confused with the personal beliefs of common people, which are to be respected, and not challenged. On the contrary, I can only stress that scientific theories are meant to be challenged. The scientists who entertain those theories, should not consider these challenges as a personal attack, but merely as a proper application of the scientific method.

Consider the fact that until quite recently, it was thought that we lived at the centre of the universe. Everyone agreed on this, and some people still believe this. It was thought, and some people still think, that all the answers are contained in the bible. That is of course perfectly acceptable, at least for common people.

Such an attitude is *not perfectly acceptable for scientists!* This is not to imply that scientists maintain that all scientific questions are answered in the holy bible. On the contrary, they have their own books of science, which they tend to regard as "holy scripture". At least, they treat them as such. The theories put forward in certain science books are *not* to be challenged! Any student of science who questions those theories, is sure to fail the course. Any scientist who challenges those theories, commits "career suicide". As a result, both students and scientists, have learned to memorize scientific theories, rather than challenging them. It is the price to be paid for being successful!

Many years ago, it was such scientists as Kepler, Copernicus and Galileo, who devoted their lives to science. They were the first to challenge the theory that the earth was at the centre of the universe. In fact, they risked their lives in studying the night skies. If they had been caught, they would have been executed.

At a later date, it was Newton who studied the work of those scientific pioneers, and brought it all together, in his three laws of motion. As Newton stated, he had merely "stood upon the shoulders of giants".

Since the time of Newton, the state of science has regressed. Once again, scientific theories are meant to be memorized, not challenged. Anyone who dares challenge scientific theories is not allowed to earn a living, working in any field of science.

For that reason, I am appealing to the members of the public. Working together, we can make a number of major scientific breakthroughs, if only by proving the existence of these animals.

I first became convinced of the existence of these animals, at the time my mother in law recognized a picture of an elephant. She mentioned to my wife that the elephant she *remembered* was *hairy!* For that reason, the Dene refer to this animal as the "hairy elephant". Of course, that is a reference to the woolly mammoth. More on that subject later.

The main thing is that my mother in law lived almost all of her life in a log cabin. She spoke broken English, gave birth to all of her children in the mountains, and only after the birth of all of her children, somewhat later in life, did she first set foot in the place she had heard so much about. That place is called "town". That is the same person who recognized a picture of an elephant! The only explanation is that she saw a woolly mammoth in her younger days! With that in mind, I began to seriously consider the fact that the stories they were telling me were completely true!

CHAPTER 2

The "Devil Bird"

If there is one thing that terrifies the Dene Elders, it is the animal they refer to as the "Devil Bird". It is their belief that it is an evil spirit, a demon, a "bird from hell".

This belief is based on the fact that the animal is nocturnal. It spends the day light hours inside caves. It comes out of those caves after sundown, and hunts in open areas. It also tends to avoid any light.

Incidentally, I have since determined that this is the same animal which is referred to as the "Dragon", in many other parts of the world. The fact is that the Chinese have named the twelve years after twelve animals, and that includes the Dragon. That is referred to, quite reasonably, as the "Year of the Dragon". I am convinced that the Dragon exists, and very soon, will prove this.

It is also a fact that the Dragon is mentioned in the bible. As the bible was written in Africa, this means that the Dragon is in Africa, as well as Asia.

It is only in the New Testament, within Revelations, that the Dragon is identified as the "Beast", or "Satan". That is precisely the belief of the First Nations people!

It is not just the Dene Elders who believe that this animal is a spirit. It is safe to say, that all across North America, First Nation Elders can swear to the fact that these animals exist. Most of them just think that it is a spirit, and not an animal. We must respect that belief.

Countless people are aware of the belief, of the Elders, in "spirits", yet most of those people consider this to be nothing more than a "myth". They could not possibly be more mistaken! But then none of them has thought to ask the Elders to describe this "Devil Bird".

That did not stop me from asking that question, as I am not terribly shy. The answer they provided, shocked me, to the very core of my being: "It has a head like an eagle, the body is the size of a man's body, the wings are as wide as two moose hides stretched out, the flapping of the wings sounds like dry hide, and the tail is as long as a man is tall, and it ends in the shape of an arrowhead"

This was a near perfect, accurate, detailed description of a pterosaur, an animal the scientists have classified as a dinosaur, one which has been *extinct for sixty-five million years! Or not!*

A little explanation is in order. This animal is a flying reptile, and the correct scientific term is that of pterosaur. The majority of these flying reptiles have long tails, and the technical term for them is "rhamphorynchus". The flying reptiles with short tails are less common, and are classified as the true "pterodactyls".

At the moment, I am mainly concerned with proving the existence of these animals, yet the correct technical name is important. Most common people have heard of pterodactyls, but not of pterosaurs or rhamphorynchus. Bear in mind that they are commonly referred to as dragons.

The wing span of these huge flying reptiles is estimated to be between ten and fifteen metres, or thirty-five to fifty feet. This is three to four times wider than the wing span of the largest flying bird in the world, the albatross. This also gives it great lifting power. In fact, it is able to pick up deer and pack them away! Quite an accomplishment, as deer weigh around 90 kilos, or 200 pounds.

The Dene Elders are aware of this, as are no doubt the Elders all across the country. For a fact, the Dene Elders do *not* go outside after sundown. They know better! As they put it, "a big bird will pick you up and carry you away!"

In fact, they have pointed out to me, certain caves in the mountains. These caves open up onto a vertical face, otherwise known as a cliff or a bluff, or "straight up". The tops of these mountains are horizontal, or "flat", to put it in popular terminology. For that reason, I refer to these mountains as "perpendicular mountains". Such mountains are the nesting ground of the pterosaurs.

This is significant, as these mountains form the "nesting ground" of the pterosaurs. With their distinctive shape, they are easily located. The point being that it is easier to locate the nesting ground of the pterosaurs, than it is to locate the animal itself.

Incidentally, the Elders tell me that those caves lead directly to hell. They further tell me that after sundown, Satan opens the gates of hell and releases his demons, his "birds from hell"! They know this because they hear a snapping sound!

Allow me to stress the fact that I respect their beliefs, just as I respect the beliefs of all common people. Which is not to imply that I share all of those beliefs.

This stands in sharp contrast to the facts they present to me. One of those fact is that after the animals comes out of those caves, it does not just fly away. Instead, it climbs straight up that vertical face.

As for those who are skeptical—and I hope there are a great many of you!—, allow me to say that I welcome such an attitude. I just wish the scientists had that attitude!

The explanation, for the fact that these flying reptiles, pterosaurs, are able to climb straight up a cliff, can be found in the fossilized track ways of the pterosaurs! I mention this as an example of combining the wisdom of the Elders, with scientific facts.

The point is that the scientists have discovered the fossilized trackways of the pterosaurs. These trackways tell us a few things about the animals, things which the fossilized bones and eggs, of the animals, cannot tell us. They tell us that the animal sometimes walks on two legs, and sometimes walks on all fours! This is to say that the wings of the animal, double as legs! The wings are used for walking, as well as flying! The pterosaurs climb up that cliff on all fours!

I should add that, once it gets to the top of the mountain, at the point where it is flat, it no doubt opens its wings. That explains the snapping sound.

This also explains the existence of the "devil dog", as the animal is called, when it walks on all fours. Of course, it is called the "devil bird", when it walks on two legs.

This is an example of separating the facts, which the scientists present, from the theories they present, even though they present those theories as facts! The facts are just that. Facts! Their theories, such as the belief that the pterosaurs are extinct, are not based upon any scientific facts. Absence of proof is not proof of absence! Just because we have no proof that they still exist—not yet, anyway!—does not mean that they are extinct! To present a theory, based on no facts, is completely contrary to the scientific method!

Now to return to the subject of the pterosaurs.

It should be stressed that this animal is nocturnal, which is to say that it avoids the light, as much as possible. It comes out of those caves after sun down, and returns to those same caves before sun rise. That is the reason it is so rarely seen.

The entrance of those caves may, or may not, be very high off the ground. It is necessary only for those caves to be high enough that predators, such as bears and wolves, cannot climb into the caves. This is to say that four to five metres, or fifteen to twenty feet above the ground, is sufficient.

Incidentally, numerous people have asked me if these animals can be spotted on radar. To this question, I can only respond that the question is backwards. As they are the size of small planes, the real question should be: *How can they not be seen on radar?* More on that subject, later on in the article.

Now to proceed with our investigation.

These animals are reptiles, so that they reproduce by laying eggs. Further, reptiles cannot generate their own body heat. The eggs of all reptiles have to receive heat from an outside source, as otherwise they cannot hatch. The eggs of crocodiles receive their heat from rotting vegetation, for example. Turtles lay their eggs in the sand, at the edge of the water, either lakes or the ocean. The sun warms up the sand, along with the eggs. But where do the eggs of pterosaurs receive their heat?

Certainly not from inside the caves, as the temperature inside those caves is roughly a constant 13 degrees Celsius, or 57 degrees Fahrenheit. That is too cold for even the eggs of reptiles to hatch. Bear in mind that the eggs of reptiles hatch at a colder temperature than that of birds. In particular, the eggs of lizards hatch at a temperature of around 20 degrees Celsius or 70 degrees Fahrenheit. Fortunately for the flying reptiles, the day time temperatures, in early May, at least in this part of the world, frequently reaches that temperature.

Of necessity, in early May, the instinct of the females to reproduce, becomes greater than the instinct to avoid the light. For that reason, the female pterosaurs come out of the nest, choose a hill top which is reasonably bare, but not too high, build a nest and lay their eggs. They then stay outside the caves and guard the eggs. The animals are able to change their skin colour, to match their surroundings, in much the same way that a chameleon can change its skin colour. In this way, they "blend in", becoming almost invisible.

When the temperature drops during the night, they sit on the eggs. Their bodies act as insulators, so as to ensure that the heat, which the eggs have absorbed during the day time, stays within the eggs.

I should add that the the place where the female builds a nest and lays her eggs, I refer to as the *nesting site* of the pterosaur. I use the term "site", in order to distinguish it from the *nesting ground* of the animal, which is the caves in the mountains. The nesting site varies from year to year, as every predator in North America is constantly looking for those eggs. As well, bare hill tops do not stay bare for long.

Many years ago, one of the Elders in the village, found a rather distinctive egg. He said it "was the size of an ostrich egg, brown, and had a thick skin". It is significant that the eggs of all reptiles have a thick shell. Of course, he had no way of knowing that the egg which he had stumbled upon, was worth a fortune!

As the eggs of pterosaurs are the size of ostrich eggs, that gives us a place to start. It so happens that the eggs of ostriches take forty-two days to hatch. It is reasonable to assume that the eggs of pterosaurs take a similar time to hatch. If that is the case, then the eggs should hatch in mid-June.

Such a hatch places the females upon the horns of a dilemma. *Reptiles do not feed their young!* Once the eggs hatch, then the youngsters are on their own! They have got to fend for themselves! The trouble being that the bare hill tops, while providing heat from the sun, also provides for a breeze. This breeze serves to drive away the bugs. These bugs are the very thing the youngsters need, in order to survive.

I should mention that the newly hatched eggs of birds, are referred to as chicks. By contrast, the newly hatched eggs of pterosaurs are referred to as "flaplings".

As soon as the eggs begin to hatch, another instinct "kicks in". The female picks up the nest, complete with eggs, and carries it to a place where there is a great deal of food for her flaplings. That place is a swamp. The food is mainly bugs, of all varieties, although I am sure they feast mainly upon mosquitoes. Most nourishing! No doubt, they also prey upon small game, such as mice and voles, as well as fish.

Within possibly six weeks, the flaplings are strong enough to fly. This is a fact, as they have been seen flying, at the end of July. Remarkable!

To proceed. As the eggs are the size of ostrich eggs, it stands to reason that the flaplings are the size of ostrich chicks. Further, as the wings are not fully developed, the freshly hatched flaplings cannot fly, so that they are forced to run around on two legs. The problem is one of eating as much as possible, while avoiding predators. Eat and avoid being eaten! Not an easy task!

Countless predators prey upon these flaplings, including birds of prey. Not a difficult task, as they are almost completely helpless. So how to explain that a few of them survive?

Perhaps we can compare them to crocodiles, also reptiles. As soon as the eggs of crocodiles hatch, the female picks them up and carries them to the water. Then the female guards the youngsters, as best she can, for a certain time. In this way, the odd youngster survives, to adulthood. Perhaps one in a hundred.

It is reasonable to suspect that the pterosaurs also stay close to their young, protecting them, as best they can. Then, as soon as the few surviving flaplings are able to fly, to guide them to the nesting grounds, in the mountains. As yet, we have no way of knowing.

It is significant that the Dene refer to these flapllings, in their own language, as "Little People". Perfectly understandable, as they are small, and run around on two legs. They have also been seen in various other parts of the world, but are known by other names. The most common name is that of "leprechaun".

Bear in mind that these flaplings, "leprechauns", "little people", will be seen only in early summer, and only around swamps.

As I have a twisted sense of humour, I once made the mistake of mentioning to a couple of girls, that as they objected to being referred to as "chicks", they should count their blessings. They should perhaps be grateful that they are not called flaplings or leprechauns! I even performed my finest W.C. Fields imitation, of "my little leprechaun". They responded by telling what they thought of me and my little joke! The language they used left no room for any misunderstanding! Strangely enough, not everyone appreciates my sense of humour!

On a more serious note, other questions came to mind. This called for a little research, which in turn provided me with a few answers. It bothered me that such a huge predator, should choose to be nocturnal.

It is the scientific opinion that the pterosaurs evolved around two hundred forty million years ago, on a huge landmass, which they refer to as "pangea". Those same scientists refer to this landmass as a "super continent", which it was not. It amounted to the grouping together of all seven continents of the world, so that it was not one continent, but seven continents, and there was nothing "super" about it. For that reason, I refer to it as the "world landmass" of pangea.

It was on this world land mass of pangea, that the pterosaurs evolved. As the continents were all connected, the pterosaurs spread to all seven continents. This to say that they spread around the world. At least, they spread around the seven continents. They may or may not have spread to various islands, such as Hawaii or Tahiti. That remains to be seen.

Over a period of many millions of years, the world land mass of pangea split up, and the continents went their separate ways. Each continent carried the pterosaurs with them. As the continent of Antarctic drifted ever closer to the south pole, it gradually became too cold for the eggs of these reptiles to hatch. For that reason, the pterosaurs died out in the Antarctic. *The pterosaurs still exist on the other six continents!*

I can only stress that these pterosaurs still exist on Asia, Africa, Europe, Australia, North America and South America. That is a fact. It is also a fact that they are predators. They hunt in darkness, in open areas. With that huge wing span, they do not hunt in heavy timber.

All predators have a keen sense of smell, and these predators are no exception. Further, all predators are unpredictable, but are far more likely to attack, when they smell blood. For that reason, they frequently prey upon girls of child bearing age. They also prey upon children, at every opportunity, as children are easy to pick up and carry away.

Without doubt, at the time these flying reptiles first evolved, the males had large beaks and head crests. These served as "decorations", adornments to impress the females, during mating season. It is also possible that these long beaks served some practical function, such as catching fish. They almost certainly were active during the daytime, as they had no competition, aside from each other. They had no reason to avoid the light of day.

That all changed, and most dramatically, when the first birds "took to the air". Some scientists are of the opinion that archaeopteryx was the first flying bird. Others are not so sure. All are of the opinion that certain species of birds evolved the ability to fly, perhaps one hundred fifty million years ago, or 150 MYA, to use the scientific short hand. The precise details are in dispute. There can be no question that a great many species of bird began to fly, at around that time. Equally without doubt, some of those species of birds were "raptors", birds of prey.

This brings us to a term which has caused a great deal of confusion, and that term is "dinosaur". The word literally means "terrible lizard", and was first coined many years ago. It has since become a house hold name, and is deeply entrenched. That is most unfortunate, as the name is completely inaccurate. For that reason, I try to avoid it, whenever possible. It is not always possible.

The misunderstanding began about two hundred years ago. At that time, the scientists of the day discovered the fossilized remains of huge animals. They were at a loss to explain the existence of these animals, as they could find no reference to them in the bible. At that time, the bible was their only reference source. This was in the days before Darwin. So they did the best they could, and called these extinct animals dinosaurs.

I have made my position quite clear in other writings. I maintain that the animals which are commonly referred to as dinosaurs are nothing other than birds. They had feathers and they laid eggs. Certain species died out, which happens, while other species evolved and gave rise to new species, which also happens. *There was no mass extinction of dinosaurs!*

Almost all scientists maintain that there was a mass extinction of dinosaurs, and argue passionately concerning the cause of that extinction. All are mistaken, and many have argued that the flying reptiles were dinosaurs. Most scientists also maintain that there has never been a mass extinction of reptiles, while maintaining that no less than five orders of reptile have gone extinct. Of course the public is confused!

To return to the time, many years ago, when raptors first challenged the flying reptiles for control of the airways. The war was immediate and brutal. No quarter! Strictly a matter of survival of the fittest. There was no middle ground. The results were clear for all to see. As anyone who has ever eaten a traditional Christmas dinner can testify, we feast on roast turkey, a bird, and not roast pterosaur, a reptile. Clearly, the raptors won the war for control of the sky, or at least during the day time.

How is it that the raptors were able to drive the flying reptiles out of the day time sky, into the relative safety of the darkness? Was their flying ability superior to that of reptiles? To answer that question, we must examine the flying skills of birds, reptiles and bats.

We know that in order for a body to become airborne, there must be a lifting force on the wings, that offsets the downward force of gravity. That lifting force is created when a wing is curved on the top and flat on the bottom. As the animal flaps its wings, the air that travels over the curves must travel faster than the air below the wings. The difference in air speed creates an area of low pressure above the wings. The air from below the wings pushes up on the wings, which creates lift. Of course, the wings of birds have a curve on the leading edge. The flapping of the wings creates lift, and the animal becomes airborne.

Modern aircraft operate on the same principle. Flaps on the leading edge combat turbulence, which can be a problem when air speed is low and the aircraft is trying to land.

We can first examine the flight of bats. These mammals are very efficient fliers, and for good reason. The hind quarters of a bat, commonly referred to as its "legs", are attached to its wings. The flapping of the wings is powered not only by the front quarters, called the "arms", but also by the legs. When the bat flaps its wings, the legs move up and down with the arms. The result is a very strong wing stroke, which results in greater lift. The tracks left by the bat are also distinctive, as they walk on the front paws as well as the hind paws, with the hind paws splayed out, far to the side.

The recently discovered fossilized trackways of the pterosaurs show that the animal walks in a manner similar to that of a bat—sometimes on two legs and sometimes on all fours, with the hind paws spread out, clear evidence that the hind paws are attached to the wings. All four limbs of the pterosaurs are involved in the flapping of the wings. Clearly, the pterosaur is a most efficient flier, far more so than I expected. Almost all the muscles of the body are involved in the flapping of the wings, which is to say flight. It is also a fact that this animal has a special plate of light weight bone that strengthens the torso and shoulders, thus eliminating most of the muscle groups that do not contribute to flight. This gives it an advantage. The more light weight the flying animal, the easier it is for the animal to become airborne and stay airborne. The pterosaur has flaps on the leading edge of the wings to combat turbulence, thus making it more stable, just as modern aircraft have flaps. Even this is not the only flying advantage of the pterosaur. The shoulders of the animal are a marvel of aerodynamic engineering. It is very likely that this animal has an extra pivot joint between the upper end of the shoulder blade and the bony plate that stiffens the torso, which allows the shoulder to swivel. This permits the shoulder bone to swing not only up and back, but also down and forward. Such movements dramatically increase the power of the down stroke of the wing.

In addition, as cold blooded animals, reptiles need to consume far less food than birds, which are warm blooded animals. These are all to the advantage of flying reptiles.

We can contrast these advantages to that of birds, starting with the flight feathers. Such feathers are attached to the wings and only to the wings. Only the front quarters of birds are involved in flapping the wings. The hind legs of birds are dead weight, when the animal is flying.

On the other hand, as warm blooded animals, birds do not require the heat from the sun to warm up and become active. At the first crack of daylight, birds, including raptors, took to the air and began to hunt—as they do to this day! By contrast, pterosaurs, as flying reptiles, had to wait for the heat from the sun, to warm them up, before they could become airborne. The amount of time required to warm up the reptile varied, depending upon the size of the reptile, the night temperatures, the weather conditions, the time of year and numerous other variables.

The brief interval, between dawn and the time that the heat from the sun warmed up the flying reptiles, was critical. Although that interval may not have been long, it was long enough for the predatory birds, the raptors, to decimate the flying reptiles, the pterosaurs. Even though the cold blooded reptiles were far more efficient fliers and required far less food, pound for pound, the raptors still out competed them for mastery of the day time skies. It is very likely that over a period of tens of millions of years, various species of raptor evolved, and gradually drove all species of flying reptiles, pterosaurs, either to extinction, or to the relative safety of darkness. This is another way of saying that the flying reptiles became nocturnal, and this happened around the world.

This is *not* to say that the raptors have won the war, and there is no more competition between birds of prey and flying reptiles. It just means that the battle ground has changed, from the day time skies, to the night time skies. Raptors such as owls and night hawks, hunt in the darkness, as do bats, in competition with flying reptiles, pterosaurs.

As the pterosaurs gradually adapted to a life of living in darkness, they were forced to make a few adjustments. In particular, during mating season, the females were no longer deeply impressed by the long beaks and head crests of the males, if only because they could not see them. The males had to find another way to display, and they did. The males evolved the ability to glow.

As for those who find this quite remarkable, consider the fact that "fireflies" have evolved the ability to glow. These insects, otherwise known as "glow worms" or "lightning bugs", are really beetles. Light production in these beetles is due to a chemical reaction called bioluminescence. The beetle has a special light emitting organ in its lower abdomen, just as there is no doubt a similar light emitting organ in the pterosaur.

In the case of the firefly, there is an enzyme, luciferase, which acts on the luciferin in the presence of magnesium ions, ATP, and oxygen, to produce light. The light is in the visible spectrum of 520 to 680 nanometers, so that we see the light as yellow, green or pale red.

No doubt, a similar chemical reaction occurs in the pterosaur, which gives off a glow from the torso, and is clearly visible from the ground. As yet, we do not know the precise wave length of light, which is to say the colour, of the light produced. We do know that each species glows in a distinctive colour. These colours are as distinctive as the plumage on birds, and will soon be used to help identify the species of pterosaur.

It is to be hoped that many readers will be inspired to look into the chemistry involved in the production of such bioluminescent light.

It is very likely that the males acquire the ability to glow, only at the onset of sexual maturity. As yet, we have no way of knowing if the females also have the ability to glow.

The reason I say this, is that there are times when evolution will take one characteristic and use it for more than one purpose. Herds of caribou come to mind. Among such animals, both males and females grow antlers. The males grow antlers in the spring and summer, in order to display in the fall, and then drop the antlers in the winter. But then the females grow antlers in the fall and winter, in order to protect their calves, which are born in the spring, and then drop the antlers in the summer.

In much the same manner, among pterosaurs, the males use the ability to glow, in mating season, as a means of display. But then those same animals also use the glow for hunting purposes! They will fly around a huge swamp, glowing brightly. The bright light attracts clouds of insects, and those insects attract bats. Then the flying reptiles attack the flying mammals.

As yet, we have no way of knowing if the females also glow, under these circumstances. We do know that many people find these lights to be quite entertaining, and in certain communities, are used as a tourist attraction.

We also know that this glow, from the torso of the pterosaurs, is the basis of a great many Unidentified Flying Objects, or UfO's. As I have now identified many of these flying lights, it is no longer correct to refer to them as UFO's but as IFO's, Identified Flying Objects. Of course, that is not about to happen, but hopefully, it will appease many of my most vocal critics, those who have been complaining that I have not been looking for UFO's. God forbid that I should be accused of being a UFOlogist! For many years, it has been my opinion that such people are not entirely sane, to put it politely. More accurately, I have long maintained that they are completely out of their minds! I can only hope that they will adopt a more scientific approach to their research.

These lights in the night sky, so called UFO's, have been associated with the death and mutilation of livestock. A popular television show has recently stumbled upon the fact that in the mountains around the world, horses and cattle which are left outside, after sundown, are being discovered in the morning, "dead and mutilated". The fact that there is no blood around the wound sites has everyone puzzled.

The writers of the television show have not stated it quite that clearly. But then, they have no desire to be accurate. They are focused on entertaining, not educating.

For that reason, they have come up with some hare brained theory to explain this phenomenon.

The "explanation", as put forward by the announcer, is that "aliens from a distant world have travelled to planet earth", presumably at "warp speed"—whatever that is-, have "beamed" these animals up to their space craft, drawn out their blood and mutilated them, and then "beamed" them back down to planet earth. Further, as "proof" of the existence of these spacecrafts, the commentator has mentioned the existence of these lights in the night sky. He maintains that the "motion of these lights does not match the motion of any mechanical aircraft, either plane or helicopter".

At least that part is true! It is not the motion of a machine, but that of a flying reptile!

It is entirely possible that the announcer actually believes the nonsense he is spitting out!

Perhaps his days as an actor on a popular television show, in which "beaming" was common place, has caused him to lose touch with reality!

In all fairness to the writers, they have drawn attention to the fact that livestock is being killed and mutilated, in the mountains, "around the world". They did not state it quite that clearly, if only because they were hindered by their mysticism. Or perhaps their only concern was that of entertaining.

No doubt, the "lights in the night sky", the"UFO's", are seen only in mating season, in the spring. Another little detail that the writers neglected to mention!

That in no way changes the fact that livestock, which is left outside after sundown, is being found in the morning, dead and mutilated. The owners of these animals are completely puzzled, as they have clearly not been killed by humans. Nor have they been killed by predators, such as bears, wolves, grizzlies or cougars. The fact that there is no blood around the wounds, leaves no room for any misunderstanding on that point!

This lack of blood around the wound sites is of critical importance. It does *not* mean that the blood is missing! It tells us that the animal was dead when it was mutilated. The question then becomes, what killed the animal?

We can rule out blood loss, so that leaves poison. What kind of poison? It certainly did not eat the poison, so it must have inhaled the poison.

The livestock is clearly being killed by a pterosaur.

There are various local names in North America, such as Devil Bird, Thunder Bird and Jersey Devil. In most parts of the world, such as Asia and Africa, they are commonly referred to as Dragons or Fire Breathing Dragons.

This is significant, as the reference to "fire breathing", is a reference to the "cloud of smoke" which the animal releases. Except that this "smoke" is not smoke at all. It is toxic, poison gas. Severely toxic! So toxic, it is capable of killing a horse! We know this for a fact, because that is precisely the poison that kills these animals!

As the subject is so important, I will mention that after the prey animal is killed, the pterosaur generally rips the flesh off the face, the genitals, and tears out the large intestine, from the rectal orifice. As the large intestine is so nourishing, this makes complete sense. The pterosaur is not strong enough to rip apart the carcass, so it has to settle for the morsels it can gather.

The standard response of the farmer is to call the police. As he is well aware that the animal was not killed by a predator, or at least not by any predator of which he is familiar, he just naturally assumes that it was killed by humans. Such an assumption is natural, but mistaken.

As the police respond, they too are puzzled, as it was clearly not killed by any human. They note the lack of bullet wounds, foot prints, tire tracks or blood. Clearly not a police matter, as no crime has been committed. For that reason, they leave.

This does not satisfy the farmer, so frequently in America, the FBI is notified. They in turn refuse to respond, as they have seen this so often. This response may be correct, but does not solve the problem.

Technically, this is a matter for the game warden. Not that the game warden is ever notified! Even if he is, this is beyond his level of training.

The scientists should be involved, but they are careful to avoid the subject. The last thing they want to do, is prove the existence of the pterosaurs! And proving that is not difficult!

Three things have to be done. The first two are quite simple and easy, while the third is more challenging.

The first thing that has to be done, is a simple matter of drawing out a sample of blood, from the carcass, and sending it to the lab. As long as the blood is drawn out within twenty-four hours, after the death of the animal, the lab should be able to determine the poison gas that was used to kill the animal. I am certainly curious.

The second thing is also quite simple. The wounds should be swabbed, and the swabs sent to the same labs, for a DNA test. The lab will report that the DNA is that of a reptile, one which is not known to science! It stands to reason that this will prove that these animals were killed by poison gas, released by a reptile. What reptile?

The answer to that question involves getting government permits, with the goal of opening up an old logging road, close to the village in which I live. It is called the Ten Thousand Road.

Once we establish the fact that this livestock is being killed by poison gas, emitted by a reptile, one which is not known to science, then perhaps the government officials will become "sweetly reasonable", and issue the permits. Then it is a matter of raising some money to hire the machines to re-activate that road.

This calls for a little explanation. The fact is that the caves in those mountains, which contain the nesting ground of the pterosaurs, open up onto the Ten Thousand Road.

I know this for a fact, because we were logging on that Road, thirty years ago. My relatives pointed those caves out to me, and told me that after sundown, the Devil Bird comes out of those caves. They were not joking! I should have paid strict attention!

After the area was logged out, the road was deactivated. Now it has to be opened up. Then it is a simple matter of setting up cameras, at the mouth of the caves, to locate the pterosaurs, as they emerge from the caves.

It is vitally important to prove the existence of this animal. The members of the public have got to be made aware of the fact that it exists, it is a predator, it hunts in darkness, in open areas. It has a keen sense of smell, and is far more likely to attack, when it smells blood.

This helps to explain the disappearance of so many girls of child bearing age!

The Highway from Prince Rupert, on the West Coast, to the city of Edmonton, in the central interior of the province of Alberta, is known as the Highway of Tears. It has earned that name, due to the disappearance of so many people, over the years, from this stretch of the Highway.

It is not by chance that this Highway runs directly through the mountains! Nor is it by chance that most of the people who have disappeared, have been girls of child bearing age!

The sad fact is that families break up, and at that time, all too often, young girls leave home. In the cities and towns, they generally jump on a bus. Human predators are well aware of this, and consistently hunt for those young girls, at bus depots. By contrast, the girls of families who live in isolated areas, including many Reserves, tend to go hitch hiking. These girls are then frequently preyed upon by pterosaurs. They generally attack because they smell the blood!

Of course, it is not just on highways that people are attacked by these flying reptiles. They hunt in all open areas, in the darkness. The areas have to be open, because the animal has a huge wing span, and the area has to be dark, because the animal tends to avoid the light. On the other hand, there are indications that the animal is becoming "braver", more familiar with artificial lights, so that such lights do not offer full protection, as was once the case.

I have been asked what to do, if caught outside, after sundown, and attacked by the pterosaur. Find shelter! With that huge wing span, they are forced to hunt in open areas. If no shelter is available, hit the ground! They hunt us by wrapping their paws around our shoulders and sinking their claws in! Do not make it easy for them! If we are flat on the ground, then that cannot be done! By all means, fight them! They *may* kill you, but if you do not fight them, they *will* kill you!

I have also been asked to clarify, the "sound of flapping wings", of that animal. To say that it "sounds like dry hide" is not terribly helpful, unless you are familiar with the sound of dry hide! Most people are not!

With that in mind, perhaps it would be best to allow a girl who was attacked by a pterosaur, to describe the event, in her own words:

"It was after sun down, but not yet full dark when we first heard a whooshing sound. Really strange. Whoosh, whoosh, whoosh. Then we looked up and saw it. It attacked us and we chased it away, so we followed it".

For what it is worth, the flapping of the wings makes a distinctive "whooshing" sound.

As for those who think that the paws could not possibly be big enough to wrap around our shoulders, think again. Those who have seen the tracks that these animals leave in fresh fallen snow, can testify to that. They assure me that the paws are possibly half a meter long or 18 to 20 inches, and 6 to 8 centimetres or 2 to 3 inches wide. As they phrase it, "real long and real skinny". It is not by coincidence that these descriptions match the fossilized trackways of the pterosaurs!

After the animal comes out of the caves, in the darkness it generally selects an evergreen tree, and with its paws, pushes over the top stem of the tree. Then, with the stem of the tree horizontal, or "flat", to state it in common terms, the animal sits on that this perch. These trees are usually located at the edge of a clearing, as the animal hunts in open areas. These clearings may or may not be on the edge of a highway or road. The animal also hunts near camp grounds, play grounds, school yards, and in fact near any open area, and that includes the backyards of people. They tend to return to the same tree, night after night. As a result of this, the top stem of those trees tends to grow on an angle, or a "slant", as is commonly stated. To identify these trees is to identify the hunting grounds of the pterosaur.

As it is only the top 4 to 5 metres or 12 to 15 feet of the tree that is bent over in this manner, that provides us with a "ballpark figure", as to the weight of the animal. I estimate the animal weighs no more than 10 to 15 kilos, or 25 to 35 pounds. I refer to this as an example of using "common sense", and it is one of the methods I recommend.

I do not regard this as an alternative to the scientific method, but as part of the scientific method.

I should add the fact that as the animal prefers to sit on the top of evergreen trees, this helps to explain the reason that experienced woodsmen prefer to build their cabins within clearings. It is partly to avoid the chance of a strong wind blowing a tree down, onto the cabin, but also such woodsmen are aware of the existence of this animal. They are also aware that it likes to sit on the top stems of trees.

In the winter months especially, the woodsmen have to come out of the cabin after dark, if only to go to the back house, or wood pile. If a tall tree is close by, they have no way of knowing if a "Devil Bird" is sitting there, just waiting for them to step outside.

These animals have truly adapted to a nocturnal life style. They come out of those caves only after sun down (aside from the females, which emerge in order to build a nest and lay eggs), at a time which is generally cooler than the day time temperatures. In fact, these reptiles are able to hunt in temperatures as cold as twenty degrees below zero Celsius, or zero degrees Fahrenheit. These reptiles are able to do this, but only because they are very large.

Only the largest of reptiles are able to withstand such cold, and only for short periods of time, due to a process called gigantothermia. It is precisely the same process that allows another species of reptile, the leatherback turtle, to survive in the very cold environment of the North Atlantic.

While outside the cave, the pterosaur loses little heat, partly because a large body size leads to a small surface to volume ratio, so that the heat exchange volume remains low. Consequently, the core body temperature is slow to change, while a spherical body and a layer of fat helps a great deal.

Local people know the animal is close and on the hunt, because they can recognize the sounds that it makes. It is able to precisely imitate the sounds it hears. Around here, that includes the sounds of "dogs barking, babies crying and women screaming". Without doubt, they absolutely hear the sound of women screaming. In fact, each time they attack women, they hear the screams!

One of the girls in this village was recently attacked, by this animal. She had a camp fire going behind the house of her father. An open area! In the darkness, the animal attacked, probably because it smelled blood. She ran into the house, terrified but unharmed, by the Grace of God. Yet her ears were ringing! At the time the animal attacked, it released a very loud "whistling sound"!

This same girl called me up, very upset. She told her friends about this, and they laughed at her! First she was very nearly killed, and then her friends laughed! Adding insult to injury!

My response: "Welcome to my world!" This also explains the origin of the name "Thunder Bird".

Incidentally, closer to town, they hear the sounds of car horns and train whistles, so no doubt they are able to imitate those sounds also.

The animal is also reasonably intelligent, at least for a reptile. I was surprised to find that they have learned to associate the sound of gun shots, with food. The local boys found that out, at the time they were "spring trapping", several years ago.

At that time of year, March and April, the fur of the land dwelling animals is no longer valuable, as the warm day time temperatures adversely affects the winter coat of the animal. By contrast, the fur of water dwelling animals, such as beaver, mink and muskrat, remain in prime condition, at that time. For that reason, the local trappers hunt them, focusing mainly on beaver. Their method is to set traps for them, as well as sometimes shooting them, as they come out of the water.

The boys learned that on the days they shot beaver, they could expect "company" after sun down. By contrast, on the days they merely trapped beaver, no shooting, no "Devil Bird" visit after sun down. The sound of gun fire is a "dinner bell" to this animal!

This brings us to Missing and Murdered Indigenous Women, MMIW. This is a painful subject, thought to be a national disgrace, as indeed it is. Which is not to say that the title is entirely accurate, as it is not only Indigenous women who are disappearing. Members of other ethnic groups are also disappearing, as well as children.

As previously mentioned, most of these people are females of child bearing age, which the animal attacks, because it smells blood. Then too, children are easier to carry away. I mention this again, as it is so important. People must be warned!

No doubt, some of these disappearances are due to human predators. Yet a great many of those that disappear from open areas, in the darkness, have been killed by pterosaurs.

Across the country, various police departments are under extreme pressure. It is assumed that there are possibly several "serial killers", men who are responsible for the killing of these girls. There is even an RCMP "Highway of Tears" task force.

The police cannot be faulted for failing to prove the existence of this animal. It is up to the scientists to prove they exist. I stress, it is the *duty of scientists!* The scientists are negligent in doing *their duty!*

It may be objected that I am being too harsh with the scientists. Some people have even gone so far as to say that the scientists may not be aware of the existence of the pterosaurs. To such kind, sensitive, tender hearted souls, I can only respond: *How can they not be aware of their existence?*

Even the most passionate, starry eyed optimists, should agree that it is the duty of scientists to explain that which puzzles people. Countless people are puzzled by UOF's, including "cigar shaped UFO's", and "Flying Saucers". As well, the "death and mutilation" of livestock, on farms, is widely known. Yet the scientists remain silent!

As the scientists refuse to perform their duty, it is up to common people, to do their duty for them. Granted, that is not fair. That is not the way it should be. It is what it is!

On the other hand, it is not terribly difficult. A simple blood test and a DNA swab. Acquiring permits to open up an old logging road is more difficult. Then there is the matter of raising money to hire the machines. But it can be done.

To readers of this book, I can only say: I am counting on you!

CHAPTER 3

Mass Extinction of Mega Fauna

I t is the scientific opinion, that possibly ten thousand years ago, at the end of the last ice age, there was a "mass extinction of mega fauna", in that five species of huge animals dropped dead, as they were unable to handle climate change. These five species include the woolly mammoth, the Jefferson ground sloth, the dire wolf, the sabre toothed cat, and the short faced bear. In fact, all of these animals are truly huge, or "mega". What is more, they are not extinct.

For the benefit of those who are not terribly familiar with scientific terms, I will mention that "mega" means huge, while "fauna" means animals. I will also mention that this theory, is "mega ridiculous"!

As all scientists are supremely well aware, within the last hundred thousand years, we have had no less than three ice ages, here in North America, all of which the "mega fauna" survived. Those same scientists are also well aware that each time the climate changed, which is to say that each time the continent cooled off and glaciers covered the land, these species survived. That merely stands to reason. It also stands to reason that each time the continent warmed up and the glaciers melted, the species also survived. The fact that these species were alive at the end of the last ice age proves this, beyond any shadow of a doubt. It also proves, also beyond any shadow of a doubt, that these species are quite capable of handling climate change. Further, they did. Those species are still very much alive.

As mentioned previously in this article, my mother in law recognized a picture of an elephant. The only difference is that the "elephant" she *remembered,* was "hairy". That is the reason the Dene refer to this animal, the woolly mammoth, as the "hairy elephant". This is to say that the *woolly mammoth still exists!* The Dene were running from the woolly mammoth in the late twentieth century! Their one and only chance against this animal, the largest land dwelling animal in the world, was to seek shelter in the safety of the nearest swamp!

As the swamp represents muskeg, and the animal is able to sense that it has to avoid muskeg, the people were safe in the swamp. In winter, this is not an issue, as the mammoth spends the winter in caves.

This in *not* to say that the mammoth is a flesh eater, because it is not. It is to say that the mammoth is almost certainly an intelligent animal, and as such, is very likely capable of emotion. The emotion it feels towards us, is that of hatred! Who can blame it? Until very recently, the mammoth was widespread across North America. It was especially at home on the prairie. Then, at the time of the European invasion, settlers appeared. The last thing a settler wants on his homestead, is a five-ton vegetarian. They made this quite clear by shooting every mammoth they could get in their sights.

The modern day farmers of Africa are currently in a similar situation. Their only livelihood is their crops, and the elephants take great delight in feasting and trampling those crops, which the farmers have worked so hard to grow. But then those crops are generally far superior, far more nourishing, than the vegetation which grows naturally in the area. No doubt, the farmers who are suffering the loss of their crops, would love nothing more than to kill every elephant which comes close to their farm. They are unable to do so, due to laws. The elephants are most valuable, if only for purposes of attracting tourists.

By contrast, the American homesteader of the nineteenth century, was not restricted by such laws. As the railroads pushed west, from the province of Ontario, they carried settlers who cultivated the land and planted their crops. They also protected those crops, with firearms, when necessary. There was no law against killing the woolly mammoth.

Those animals responded by running ever further west and north, into the relative safety of the mountains. The remnants of a once great herd, which at one time stretched across North America, is now scratching out a living, in the unforgiving mountains.

The immediate problem now, is to prove they exist. Then it will be a matter of passing laws to protect them. At that point, they will have to be led out of the mountains, to their former grazing grounds. Wild life photographers will be anxious to take their pictures!

The farmers on the prairie will not be so enthusiastic! There is a reason their ancestors shot those animals! To protect their crops! Yet soon, they will find themselves in the same position as the African farmers!

The mammoth will find themselves in a land transformed, a "land of milk and honey". They will marvel at fields of wheat, rye, corn and oats, as far as the eye can see! The mammoth can see very far! They will then walk through the fences of the farmers, and have a feast! The difference now is that the farmers will not be allowed to kill the mammoths! They will be protected by law, just as the elephants of Africa are protected.

The government will have to find some way to compensate the farmers, for the loss of their crops, perhaps by charging a fee to the tourists, those who delight in taking pictures of those magnificent animals.

No doubt, in the fall, instincts will kick in, which in turn will force the mammoth to migrate south to warmer pastures, in the process, passing through international borders.

Once the mammoth gets the idea that it is safe, that we are not going to shoot it, it may decide to "get even". It may well do its best to irritate us. That could include blocking highways, trampling golf courses, knocking down fences, and anything else that crosses its mind. We had best be prepared.

As the animal is so huge, it should not be terribly difficult to locate. It can certainly be seen from small planes. The pilots who fly around these mountains, referred to as "bush pilots", no doubt see this animal on a regular basis. Equally without doubt, certain forest service personnel, those who accompany the bush pilots on fire patrols, in the forest fire season, also see those mammoths. They are careful to report the location of all forest fires, but are equally careful to not report the location of any mammoth. Their careers depend upon remaining silent!

This brings us to another species of "mega fauna", also allegedly extinct. This animal is commonly referred to as the "short faced" bear, technically referred to as "arctodus simus". The Dene refer to it as the "rubber faced" bear, or the "beaver eating" bear.

This bear is truly huge, weighing in at a ton. This is to say one thousand kilos, or two thousand pounds. It stands one point eight metres or six feet at the shoulder. When it rears up on its hind legs, it stands five metres or seventeen feet high! The claws are seven inches or fifteen centimetres long. They use these claws to *tear apart beaver houses!*

As far as I am aware, no one has ever documented a bear tearing apart a beaver house. For that matter, I have never heard of anyone, witnessing a bear trying to tear apart a bearer house, if only because they know that they cannot do this! Yet this bear makes a habit of tearing apart beaver houses! It is that big and powerful!

This is to say that in spring, when the beaver pond is still frozen, the short faced bear tears open the beaver houses. Of course the beaver dive into the water, but they still have to come to the surface to breath. With the pond frozen, the only opening is into the beaver house, and the jaws of the short faced bear.

To clarify, the Dene tell me that the bear has no hair on its face, which is the reason they call it the rubber face bear. The reason they call it the beaver eating bear is quite obvious.

I have included two stories of encounters with the short faced bear, by a Dene Elder, Seymour Isaac. He gave me his permission to include them in this book. I should add that these incidents happened at a time when he and his brother Francis were lads. The men were teaching them how to "spring trap", so the boys very likely had twenty-two calibre rifles. The men had high powered rifles, or "big guns", as they refer to them.

I have chosen to include those stories, as they were written by Elder Seymour, since I consider them to be so important.

But now allow the Elder to tell his story, in his words. The title is:

Beaver Eating Bears

I remember the first beaver eating bear that was shot and killed, way back in 1953 or 1954. Grandpa Keom Pierre and his stepson, Larry Pierre and William Poole and Jimmy Dennis, Charlie and I, who were all young boys at the time, not yet in our teens, were trapping beaver. We had been doing so for five days and headed up to Shovel Creek Pass on the far side of the trap line which Grandpa Keom called 15 Mile Cabin. Around the first week of April, there was still four or five feet of snow, so we used snowshoes. We had set beaver traps under the ice and found some places that had exposed spring water.

We knew that Davis Creek flowed past 15 Mile Cabin only about three miles away. Keom and Larry also knew that the Davis always opened up early, so the boys were told to bed early. At first light, we would head to Davis Creek while the snow crust was still frozen. The next morning, we all got ready to travel on top of the crusty snow and took three of Grandpa's hunting dogs with us. The dogs were named Tez, Pup and Silver. We travelled on and finally came to the creek. The dogs started barking a commotion like they had seen something they did not like and they certainly did. If front of us stood the biggest bear, we had ever seen; it was bigger than the biggest bull moose we had ever seen. Then the fun began.

They started shooting and Keom just shouted, "Don't hit the dogs", but the dogs had the advantage because the bigger bear was breaking through the four-inch snow crust and the dogs were on top of the snow and not sinking through. Since the bear was breaking through the crust, it could not move fast. Once that bear saw us, the bears concentration was on the dogs. The bear tried to get them, and the men tried to move away from the bear as they were shooting. After twelve or thirteen well placed shots, the bear finally dropped. Grandpa Keom had used an 8 millimetre war gun, Uncle Larry had used a 303 British war gun, and Jimmy fired his brand new 30-30 Marlin lever action. But it was William, with a single shot .22, who took the beaver down that day.

While the beaver was being shot at, it thrashed everything and anything that was in its way. It ripped out little trees; willows flew left and right. More than a few times, the bear stood up and was a massive seventeen feet tall. Man, did he stink! After the chaos was over, the bear was skinned. Its claws were six to seven inches long. The hide could not be saved because the side he slept on was ruined from his shoulders to his hips from rubbing.

Here is the second story from this Elder:

Beaver Eating Bears In Akie

In 1958, my brother Francis Isaac and I were up the Akie River to do some spring beaver trapping with Uncle Angus Pierre and Uncle Mac Pierre and the whole family. We just loved Grandma Elizabeth and the great stories that she always told.

At about mid-May, the five of us being myself, Uncle Angus, Auntie Lucy, Uncle Mac and my brother were going up the Akie. Old Grandma was left to babysit at the time.

We started to travel up the river. As I recall, the Akie is a very difficult river to shoot beaver on in the spring. At that time of year, the river is so low it almost becomes a creek. When the river is low, the beavers do not come out until dark, so we mostly used traps. We did pretty good in trapping on a daily basis and trapped three or four beaver each day. When we baited the traps, we used poplar tops mixed with beaver castor and used a little of our own oil mix, which worked very well in the traps as bait. Since we camped along the way, we would set our traps and enjoy ourselves until it was time to check our traps.

As we cut trail along the way, Uncle Mac told us to look out for ourselves as there was a lot of bears and grizzlies in the area, especially where we were cutting trail. During our way along the trails, we encountered bears and scared them away. We didn't encounter any problems until that one fine day on the way back. We passed the second trapping cabin and went right on up to the area that's called the big bend in the Akie River.

Our Sekani ancestors had a name for that place and called it Big Bad Yorks. Uncle Mac shared some history of that place. Our people used to stop there to dry meat, and the women picked berries and medicinal plants. Big Bad Yorks was a nice place to camp. It got its name by being a place where most of the valleys could be seen. Creeks ran into the Akie, and there was plenty of game in that area. This place should be marked as a historical marker on our map as a secret place.

We had camped in the area for about three days before heading down the Akie. On our third day of camping, we encountered what our ancestors called the beaver hunter.

At the time, Uncle Mac was setting a beaver trap. Uncle Angus told him that we would travel on and wait down at the point for him. We came to a place with a lone sandbar right by the bluff. The water must have been quite deep as I remember an eddy in the green, pristine water.

In that area, we saw a beaver jump in. Uncle Angus took out his gun and went down that bluff to look for that beaver. Right across the river, we saw a dry slough that ran right into the Akie. I thought we had seen a moose and Francis thought that too. All of a sudden, Francis looked at Uncle Angus and wanted to know what was on the sandbar? At first, we thought it was maybe someone, but it was a beaver hunting bear coming towards us on the sandbar. It was only fifty feet away from us. Uncle Angus was ready, and grabbed his gun and shot. Those first two shots were so fast and accurate that it sounded like one shot! Uncle Angus shot four more shots then ran and knelt by a tree. He started shooting more. Although Uncle Angus had a single shot 30-30, he was shooting shell like an automatic. Oh, the action! In the meantime, when all this shooting was going on, we could hear that beaver hunting bear yelling and hollering, crawling, rolling around, and scooping up paws full of gravel. The huge bear stood up. After fifteen shots, he had to have felt the shots and started walking on the island. One 30-30 and one 303 British has hit him. Finally, we heard his last, big growl. Some of us wanted to check the bear out, but Uncle Angus noted it was getting late, and we knew it was dead.

The next day, we got up early and headed back home. We told old Grandma and the kids about the encounter with the beaver eating bear. She told us she knew about it. As long as everyone was okay, she was happy.

Between the four of us, in twelve days we got forty-five beaver and twelve muskrats!

After we were back, Billy and Art Van Somers came up and brought supplies. They brought a parcel and a letter for me from Dad saying he wanted both of us back to town. He also mentioned he had missed us very much and that he was staying with our sister Louise and her husband Wilson in Summit Lake. When they came back, we went to town with them and shared our proud hunting story.

Now to continue with our list of mega fauna, the Jefferson ground sloth is named in honour of Thomas Jefferson, the third president of the United States. The Dene refer to this animal as the "giant beaver". It is also not extinct.

In all fairness to Thomas Jefferson, the man was a genius, truly talented. He was a diplomat as well as a politician, a scholar, inventor, scientist and architect. He personally designed the building on his plantation of Monticello. It is considered to be an absolute masterpiece. This is the man who first described the animal which is named in his honour, the Jefferson ground sloth.

This in no way changes the fact that the man was a slave owner, a true psychopath. Nor does it change the fact that the scientists are supremely well aware that the animal was alive two hundred years ago, as it is today. Yet to this day, the scientists insist that the animal is extinct!

Another example of the hypocrisy of the scientists, is the supposed extinction of the dire wolf. The wolf is huge, the size of a deer. It may weigh perhaps one hundred fifty pounds or seventy kilos. The Dene refer to this animal as the "wilderness wolf". It is well named, indeed "dire", as this wolf is not afraid of us! They come into the village after sundown and prey upon dogs. Those dogs which are tied up, are particularly easy prey animals.

Remarkably enough, even the scientists admit that the dire wolf still exists, even though they claim that it is extinct. As they phrase it, the dire wolf exists, but only as a "remnant population of an extinct species". Such nonsense is now referred to as "alternative facts", although I have another name for such foolishness. I call it what it is. A pack of lies!

This brings us to the sabre toothed cat, technically referred to as smilodon fatalis. This huge cat, the size of a Siberian tiger, loves the grass land, so is not located in the mountains. It lives in the prairie, and in fact has been seen on numerous occasions, within the city limits of Milwaukee. For that reason, it is commonly referred to as the Milwaukee lion. There are numerous pictures and videos of the animal, posted on the internet. It is a female, which explains the reason the animal lacks huge teeth. Such teeth are characteristic of the males of the species only. The males have evolved those "sabre teeth" for purposes of display.

CHAPTER 4

Lake Monsters and Sea Monsters

Countless members of the public are fascinated with the animal known as the "Loch Ness Monster". This despite the fact that the scientists dismiss all sightings, as "products of an overly active imagination", at best. At worst, they refer to these sightings as "deliberate deception".

The scientists could not possibly be more mistaken! They have chosen to disregard solid evidence, which supports the eye witness accounts.

In particular, reports of a "monster" inhabiting Loch Ness, "date back to ancient times". There is even a famous stone carving, by the Pict, depicting a "mysterious creature with fins". As it was carved by people many hundreds of years ago, it is strong evidence that the animal existed, at that time. Further, as those same people went to the considerable effort to carve the symbol in stone, they must have held it in great regard.

There is even a more recent account dating from the year 565 CE, concerning St. Columba. The scientists dismiss all this as "Scottish folklore".

More recently, in 1933, a couple spotted a "dragon or pre historic monster", which "disappeared into the water", as was reported in a local newspaper. As these reports seemed to be credible, a big game hunter investigated and found "large footprints", along the shore of the lake. But then zoologists at the Natural History Museum dismissed these tracks as a hoax.

To think that key evidence was dismissed out of hand!

Shortly afterwards, a famous photograph was published. This gave birth to speculation that it was a marine reptile, a plesiosaur, one which supposedly went extinct, millions of years ago. This picture is now regarded as a hoax.

Yet there have been a great many eye witness accounts, by people who are completely trust worthy. This has given rise to the actions of a number of people, who have chosen to conduct a serious search for this animal. This has involved boats with submersibles and under water cameras, as well as sonar. All have turned up empty handed, and with good reason. They are looking in the wrong place, at the wrong time!

The Loch Ness Monster is not a reptile, a plesiosaur, but a mammal, a whale, by the name of basilosaurus. This whale has legs. It is a walking whale. It is also nocturnal. It spends the day light hours inside caves. It also spends the winter months inside those caves. It is a predator, and as such, has a keen sense of smell. It is far more likely to attack when it smells blood. That helps to explain the occasional day light sightings of this animal.

Yet it stands to reason, that a fresh water lake cannot possibly support a population of whales. There is simply not enough nourishment in the water! It follows that it is not the lake which supports these whales, but the *ecosystem!*

Basilosaurus is a predator, but not a carnivore. It is an *omnivore!* It eats flesh and it eats vegetation! After sun down, it comes out of the water and *grazes!* No doubt, it mainly eats grass, although it very likely also consumes other vegetable matter. That explains the tracks along the lake shore!

Without doubt, the huge, swimming animal in Okanagan Lake, here in North America, is also basilosaurus. There have been numerous eye witness accounts, all of which are credible. Everyone agrees that it cannot possibly be a reptile, as it "does not swim like a reptile". They also agree that it is very long, "snake like". In fact, basilosaurus is twenty metres long! They were not exaggerating!

Even though the animal in Okanagan Lake, referred to as "Ogopogo", is the most famous, here in North America, it is by no means the only "lake monster". There have also been sightings, in various other lakes, including each and every one of the Great Lakes. It is reasonable to assume that they have "taken up residence", in all large lakes, here in North America.

As previously mentioned, all efforts to locate these animals have failed, both here and in Britain. Yet very soon they will be proven to exist. There is no need for boats, submersibles, under water cameras or sonar. A simple, motion activated trail camera is all we need. It is a simple matter of attaching the trail camera to a tree, on the edge of the meadow. As the whale grazes, eating grass in the moonlight, no doubt the camera will take a picture of the animal.

The people who live around those lakes, frequently hear the sounds the animals make, as it grazes. By all means, work with those people. In this manner, common people can take part in a scientific breakthrough. We do not need a degree in science, in order to work in science! As the scientists are determined to *not* perform their duty, then it is up to common people to perform that duty!

That will resolve one mystery, that of Ogopogo and the Loch Ness Monster, as they are the same animal. Yet the confusion created by various references to "plesiosaurs" will remain. As I consider it my duty to clarify the confusion created by the scientists, I have chosen to explain a few things.

Let me start by saying that, as previously mentioned, the scientists are convinced that no less than five orders of reptile have gone extinct, even though they also maintain that *there are no mass extinctions of reptiles!* Further, it is their opinion that four of those orders of "extinct reptiles", include the long necked plesiosaur, the short necked plesiosaur, the mosasaur and the ichthyosaur. Swimming reptiles, one and all. No reason is given for this "mass extinction"!

As is well known, it has long been my contention that this "mass extinction of reptiles" is nothing more than a scientific fairy tale! I have documented that most of the animals which have been classified as "dinosaurs", were nothing other than birds, while others, such as pterosaurs, are flying reptiles. I also maintain that the animals they refer to as "great sea lizards and snake necked plesiosaurs", are also still alive. They are just not in the fresh water lakes.

It stands to reason, that they must be salt water reptiles. They too, must be located.

With that in mind, some of the techniques I suggested, in other writings, can now be applied to the coastal regions. As these animals are reptiles, they must come to the surface to breath. So that rules out the "deep blue sea"! Yet there are few reports, of which I am aware, from sailors, concerning these huge animals. This suggests that these animals are also nocturnal, perhaps spending the day light hours in caves, along the coast. It follows that they must stay close to land.

The males may have also evolved the same method of display, as that of their flying brethren, the pterosaurs. In other words, these swimming reptiles may be able to glow! If so, then that explains the underwater lights, which have so mystified people.

In fact, there are numerous reports of underwater lights. The trouble is that the details, which are of such vital importance, are quite scarce.

Among those few "rare gems", are the fact that these lights have been spotted "off the coast of California", as well as "close to the Soloman Islands". That is not much to go on, but it is a place to start.

It stands to reason that the caves, where they find shelter, must be on the coast, close to the location of the lights. It further stands to reason that these animals, as reptiles, lay their eggs in the sand of the beach. Equally without doubt, the flying predatory birds, referred to as raptors, birds of prey, are also well aware of the location of this nesting site, and of the time of the year that the eggs hatch. After all, the survival of all predators depends upon their being able to take advantage of all food sources.

We can use this to our advantage. Rather than looking for the nesting ground of these reptiles, we can instead look for a gathering of raptors, next to the beach. As they sit on the branches of trees, in anticipation of a feast, and there are great flocks of them, they are easy to spot.

Of course, the coast of California is vast, and it would be nice to narrow our search area, but it may not be necessary. It is quite possible that these animals are widespread.

My suggestion is that the newly created Councils, or at least those which have taken shape close to the coast, should assign teams of working people to investigate. Each team should be assigned a certain stretch of the coast. Their assignment should be to determine if anyone living along that stretch of the coast, has seen any such gathering of raptors.

As such coastal areas tend to be densely populated, it is very likely that the people who live there, have noticed such gatherings.

I must stress that this quest, for pre historic animals which the scientist claim to be extinct, is to be considered *part* of the class struggle, and not a substitute for that struggle.

As soon as the people determine the time of the gathering of raptors, then it is a simple matter of joining the birds of prey, as they gather. Without doubt, the eggs of those swimming reptiles are about to hatch. At they hatch, and make a dash for the water, grab one or two. A living, breathing reptile, which is thought to be extinct, should convince the most skeptical! Mission accomplished!

CHAPTER 5

Sasquatch or Bigfoot: A Separate Species of Human!

All scientists are agreed, that of all the species of human that have ever evolved, we are the one and only species of human still living. In this too, they are mistaken. Here too, all reports, many of which are highly reliable, of giant, naked, hairy humans, are carefully ignored. As well, countless plaster casts of footprints are dismissed as fakes.

I maintain that there is a separate species of human, currently walking the earth. Further, they live among us, here in North America. That species is technically referred to as Gigantopithecus, although they are most commonly referred to as Sasquatch or Bigfoot. The Dene refer to them as "Stink People", a title which is less than flattering. I refer to them as Giants, because they are truly huge.

The scientists believe that this species of ape evolved many years ago, in Asia. As it stood ten feet tall, or three metres, and weighed twelve hundred pounds, or five hundred kilos, it was by far, the largest ape ever to walk the earth. In fact, it was twice the size of a gorilla! But then, the scientists maintain that around one hundred thousand years ago, this huge ape dropped dead, for reasons which no one can imagine.

The scientists are mistaken. Gigantopithecus did not go extinct. They first became bipedal, which is to say that they started to walk

46

on two legs, and then they developed the opposable thumb. This is to say that they can now touch their fingertips, with their thumbs. In other words, they evolved into humans.

It is my contention that all apes, which first became bipedal and then evolved the opposable thumb, are nothing other than a species of human. As humans, they then began to bury their dead. That is the reason no recent remains of these apes has been discovered.

It should come as no great surprise to anyone, to find out that almost all scientists disagree with me. At least, in public!

As members of a separate species of human, these Giants deserve our utmost respect. They have every right to live their lives, as they see fit. We have no right to interfere in their way of life. We certainly have no right to hunt them, as so many people are hunting them, as if they were vermin. Such a killing can be seen as nothing less than an act of murder, completely unjustified. The fact that many of the people who are hunting them, are doing so with good intentions, does not change that fact. Certain people think that this is the only way to prove they exist! The end does not justify the means!

I maintain that these Giants are members of a hunting-gathering society, so that they are constantly travelling. To stay for any considerable length of time in one area, would quickly result in the exhaustion of the food supply, if nothing else. For that reason, they are constantly moving, following the herds. As well, the wild plants are harvested in the proper season, in the proper location.

In the spring, they travel north, in order to take advantage of the abundant supply of eggs, as the birds are nesting, at that time. As well, the moose, deer, elk and caribou give birth, at that time. They also harvest the wild berries, as they ripen, over the course of the summer. In the fall, they travel south, in order to take advantage of the warmer temperatures, as well as different food supplies.

The further south they travel, the more settlements they pass through, in the darkness. The people who live in those communities, take note of the tracks they leave behind. They know full well that something big had placed those tracks, the previous night. At the same time, they generally overlook the fact that, at the time the tracks were being placed, *the dogs were not barking!*

This silence of the dogs is significant, as dogs are known to bark at all animals. At least, all animals known to science. The fact that the dogs are terrified, to the point of silence, is due to the sheer terror, inspired by these huge people. If nothing else, this alone is reason for giving the Giants our utmost respect.

As many people are hunting them, including those who are chasing them around the mountains, their lives are being made quite miserable. They are constantly on the run. This constant harassment can result in something more than inconvenience. As they are members of a hunting-gathering society, their food supply is completely unstable. It varies according to the season, and changes in the weather. They have no emergency food supply to fall back upon. They live constantly on the edge of starvation. Hunger is a constant companion.

In addition, with our constant expansion and industrial development, we are destroying their hunting and gathering grounds. This industrial development includes dam building, logging, mining and road building. This makes it difficult for the young males to travel to different groups of people, to whom they are not related, in order to find wives. This is the one and only way to avoid inbreeding. Bear in mind that inbreeding can lead to the extinction of the species.

That is not an exaggeration. The current European nobility can testify to the fact, that inbreeding can lead to disaster! All of them are related, and their offspring, the heirs to the throne, can at best charitably be referred to as "simple souls". Anyone who doubts this has only to face the fact that the queen of England and her husband are closely related.

Ever since the birth of their first child, they have no doubt regretted the fact that they "tied the knot". The heir to the throne of England, is a constant source of embarrassment to the Queen. For that reason, it is now fashionable for the nobility to insist that their young people marry "commoners". This is not to say that they find us any less contemptible, because they do not. They have merely been forced into this.

The "royal watchers" are now speculating that the queen may have already decided, that the next king of England will not be her first born male child, but her first born male grandchild. That child is not as deeply inbred as his father!

The point I am trying to drive home, is the fact that inbreeding is certain to lead to disaster. We have no way of knowing the size of the population of the Giants, but we do know, that as members of a hunting-gathering society, their numbers are limited by their food supply. This is another way of saying that there cannot be a great many of them. If their population is reduced to a very low level, or if the males are unable to travel to other groups, in search of wives, then the species is certain to go extinct.

To be responsible for the extinction of another species of human, goes beyond genocide. Yet unless we change our ways, that could happen.

For that reason, it is urgent that we prove the existence of these Giants. Many people have been attempting this, but they are going about it in the wrong way! A different approach is required. Instead of chasing them, attract them! Allow them to come to us. They have no reason to trust us, as we are quite fond of shooting them. It will take a while to convince them that we mean them no harm.

The only members of our species that they trust, is the Indigenous people. The Giants and the Indigenous people respect each other. They leave each other strictly alone. Live and let live. The Giants can be found on the Reserves, along the Pacific West Coast, on the beaches, after sun down. The Indigenous people enjoy the beach during the day, while the Giants enjoy the beach after sun down. It is on the beaches of these Reserves that we can meet them.

The key to this historic meeting, between two species of human, lies with the Indigenous Elders. They have about as much reason to trust us, as do the Giants. For that reason, I can only suggest calling the Band Office, of a Reserve on the Coast. Request a meeting of the Elders Society. Assuming they agree, then arrange a feast, for the Elders. Be sure to bring in food those people love. As well, gifts such as tobacco, Hudson Bay blankets, electric blankets, dried fish and tanned hides. The management can advise the gifts, which the people appreciate. Spare no expense, as we are looking at a major scientific breakthrough. Make sure that food and gifts are taken to the Elders who are too feeble to attend the meeting. Respect is essential to the success of this project.

With that in mind, offer to hire a translator. It is very likely that all the Elders understand English, but a little show of respect goes a long way.

After the Elders have eaten and received their gifts, explain to them that we are interested only in meeting the Giants. We mean them no harm. On the contrary, there are countless people who are interested in killing them. The one and only way of preventing this, is by having laws passed to protect them. Further, the only way this is about to happen, is by first proving they exist. That is where the Elders come in. It cannot be done without the help of the Elders.

With that in mind, mention that the ideal place to make contact with the Giants, is on the beaches of the Reserve. This can only happen with the blessings of the Elders, and the cooperation of the young Indigenous people. The Elders know the time the Giants are on the beach, and with the help of the young people, gifts can be placed on the beach, close to sun down, for the Giants.

As the Giants are human, there can be no doubt that they love the same things that we love. That includes meat, fish, vegetables and fruit, raw as well as cooked. It also includes cosmetics and mirrors, preferably small metallic mirrors. Decorations of cosmetics, metallic ornaments, tied together with dental floss, is also an idea, as well as burlap bags.

As for those who complain that these are merely "beads and trinkets", I can only respond that you are absolutely correct. May I suggest that our goal is to meet these people, to prove that they exist. We want to impress upon them, the fact that our intentions are honest, and gifts of platinum, gold and silver, are not about to do a world of good. They have no use for such items! On the other hand, no doubt all of them want to know what they look like! All members of our species are curious! Just as members of our species enjoy wearing cosmetics, no doubt the Giants will also! As the Giants are just as human as we are, it is reasonable to assume that they love the same things that we love.

Perhaps it would be best if government officials are not involved in this science project. Such officials tend to have their own agenda, which is frequently at odds with the goal of the project. At best, such officials merely spread confusion. At worst, they sabotage the operation. It is a mistake to expect anything better from such people.

Then again, it would be so nice if even one of them was to surprise me, by doing something intelligent!

It is only after we prove the existence of these people, the Giants, that we can focus on getting laws passed to protect them. This can only be done through the United Nations, as the Giants have no knowledge or concept of international borders. Besides, the local authorities have other priorities.

Of course, once we prove the existence of the Giants, as well as numerous other species. we can expect the scientists to take credit for this. They do that, as an absolute last resort. They are quite predictable!

CHAPTER 6

Monopoly Capitalism and Our Civilization In Decline

I t is remarkable to think that for over thirty years, I have been working on this, my "little science project", as I refer to it. During that time, I have been challenging various scientific theories. This includes, but is not limited to, the theories of the mass extinctions of numerous orders of animals. The simplest way to disprove those theories, is by proving the existence of those same animals. Which is not to say that simple is easy!

For the benefit of those without a scientific background, I should add that an "order of animals" is defined as "a taxonomic rank used in classifying organisms, comprised of families sharing a set of similar nature or character".

It may help to think of the pterosaurs as an "order" of flying reptiles, composed of numerous species, located in different parts of the world.

There are currently a great many people who speak nostalgically of the time, before the age of Newton and Darwin, as the "good old days". It was certainly a much simpler time!

It was accepted, and in fact it was insisted, that the world was at the centre of the universe!

The answers to all questions, were to be found in the bible. Anyone who challenged this belief, could be charged with heresy and burned at the stake. It was this threat of execution, that served to limit scientific advancement.

All of that changed with Newton, and his three laws of motion. We now know that the earth is anything but the centre of the universe!

Other scientific break throughs followed. In particular, Darwin did for biology, that which Newton did for physics. Not everyone was terribly happy about this. To this day, there is no shortage of people who resist change! As well, there are a great many people who object to the theory of evolution, based on religious beliefs.

In particular, those who are referred to as "fundamentalists", or "creationists", deeply disapprove of the theory of evolution.

The beliefs of these people, common people, members of the public, must be respected. It is their religion. To even suggest that apes and humans have a common ancestor, is to insult them!

That being said, it is important to separate science and religion. Religious beliefs have no place in the class room! To teach "creation science" in the schools, as an "alternative" to the theory of evolution, cannot be allowed. After all, "creation" is a belief, not a scientific theory.

As that is the case, it is essential that we distinguish the fundamentalists, from the reactionaries. The fundamentalist has their religious beliefs, while the reactionaries are those who want "everything to stay exactly the way it is". Such people are opposed to any social or political reform.

As for those who are of the opinion that such people are rather quaint, harmless sorts, you are mistaken. Feel free to face the facts.

Over the last several thousand years, numerous civilizations have come into existence, flourished, risen to a peak, and then fallen into decline. This decline in civilization is not an Act of God. It is an act of people! Honest, hardworking people, build up a civilization. As opposed to this, are the reactionaries. They are lying, thieving, destructive people, intent only on tearing down anything that has been created.

Under our present civilization, our ancestors have worked all their lives to build a better world. They made great sacrifices for us, their descendants. Some of them made the "ultimate sacrifice", while others worked themselves into an early grave. Without doubt, they built this civilization. Equally without doubt, it has now passed its peak, and is in decline.

I can only stress that this decline is an act of people, reactionaries, not an Act of God. Unless we take acton now, our civilization will go the way of all previous civilizations, the "way of the dodo bird". The reactionaries would have us believe that we are "destined" to fall into decline, to join all previous civilizations. Such is hardly the case!

Our civilization is exceptional, in that we have experienced an industrial revolution! In the process, two new classes were created. The bourgeoisie, and the proletariat. Both classes have played a revolutionary role.

Marx and Engels have made that quite clear, in the Communist Manifesto:

"The bourgeoisie, historically, has played a most revolutionary part.

"The bourgeoisie, wherever it has got the upper hand, has put an end to all feudal, patriarchal, idyllic relations . . . It has resolved personal worth into exchange value . . . has stripped of its halo every occupation hitherto honoured . . . has torn away from the family its sentimental veil, and has reduced the family relation to a mere money relation".

That is "on the one hand", so to speak. The Communist Manifesto goes on to say:

"The bourgeoisie . . . has been the first to show what man's activity can bring about. It has accomplished wonders, far surpassing the Egyptian pyramids, Roman aqueducts and Gothic cathedrals . . . The bourgeoisie cannot exist without constantly revolutionizing the instruments of production . . . uninterrupted disturbance of all social conditions . . ."

The Manifesto draws a stark contrast to all "earlier epochs", before the industrial revolution, in which the "first condition of existence" was in maintaining the "old modes of production in unaltered form". This is to say that previous civilizations were determined that nothing should change.

This Manifesto was written in 1848, at a time when capitalism was still at the stage of competition.

That changed, around the beginning of the twentieth century, at which point capitalism reached the stage of monopoly, referred to as imperialism. Not too surprising, monopoly capitalism, imperialism, has characteristics which are different from those of capitalism in its early, competitive stage.

It was Lenin who subjected imperialism to a careful analysis, in his book, Imperialism, the Highest Stage of Capitalism. He stressed the point that imperialism, is "reaction, right down the line". That is another book that deserves careful consideration.

Yet the industrial revolution gave birth to a second class. That other class is the class of proletarians, the working class. The proletariat has nothing to sell but their labour power. This class too, is revolutionary. This is the class that will turn around the decline in our civilization.

I can only stress the fact that all previous civilizations fell into decline, because there was no revolutionary class to reverse that decline. None of those civilizations passed through an industrial revolution, so that no proletarian class was created.

An indication of the decline in our civilization, is to be clearly seen in science.

Formerly, scientific theories were challenged, properly so. That is the proper scientific method! That is the one and only way to determine if a theory is correct!

That stands in stark contrast to the present day, in which all too many theories are considered to be "sacred", not to be challenged! That includes the "mass extinction" of so many orders of animals, which never happened!

This retrograde trend in science is nothing other than the result of the monopoly capitalists, the imperialists, all of whom are reactionaries, interfering in science.

Mind you, they do not refer to themselves as capitalists. They refer to themselves as "entrepreneurs", or "merchants", "business people". As if changing the name changes, the nature of the beast! They live "in the past", and are determined that nothing will change! The science books have been written, and are not to be rewritten! The same is true of the history books and all text books used in the schools!

The fact that all of those books are filled with distortions and outright lies, does not impress them. The current glorification of slave owners, such as Jefferson and Washington, is merely an indication of the mind set of reactionaries.

As part of their method of rule, the capitalists have established very high rates of tuition, for university training. This has given rise to very expensive student loans. Further, any and all students who wish to pursue a career in science, must not "rock the boat". This is to say that they must not challenge the theories that are presented. The penalty for such "heresy" is "career suicide". In other words, they are not allowed to earn a living working in their chosen field. This is referred to as being "black balled". At the same time, they must still find a way to earn a living and repay the loans.

In this way, scientists are degraded. They are forced to embrace theories that they know to be false, as that is the only way to survive.

By way of exception, there are the few students who are members of families who have enough wealth, to pay for the school tuition. Such students are supremely class conscious. They know better than to "rock the boat"!

Most children of very wealthy families, usually do not pursue a career in science. Such families are referred to as "upper middle class", or "upper class", bourgeoisie. The students are generally focused on careers in business, law and politics. Such careers tend to overlap.

For that reason, most of those whom earn a degree in science, are of "middle class" background, which is to say, petty bourgeois. They also generally graduate deep in debt! This helps to explain the silence of the scientists! The only way they can pay off the huge student debt, is by working in their chosen field of science! The only way they can work in their chosen field, is by not "rocking the boat"!

As a result of this, most common people, members of the public, are blissfully unaware of the existence of these prehistoric animals, including the pterosaurs.

It is the pterosaurs, in particular, that frequently prey upon people. They hunt in open areas, in the darkness, and as predators, have a keen sense of smell. It is characteristic of predators that they are far more likely to attack when they smell blood. That is the reason it mainly preys upon girls of child bearing age. It also preys upon children, as they are easier to pick up and carry away.

No doubt, the scientists are aware of the existence of these huge species. Equally without doubt, they remain silent in order to protect their careers.

I have no career to protect, as I committed career suicide many years ago. It was the price I paid for "rocking the boat". For that reason, the capitalists cannot threaten me with career suicide. Besides, my days of "rocking the boat" are over. Now I am "blowing the boat right out of the water"! Soon, a great many careers will be ruined, or at least severely compromised, as we prove the existence of these huge animals.

As the scientists either cannot, or will not, challenge any scientific theories, it is now up to the common people to "blaze the trail". To prove the existence of these species is not difficult.

Then again, the revolutionary motion which is sweeping the world, is growing ever stronger. It is just a matter of time before the scientists are affected by this revolution. When that happens, we can expect them to rise up, band together and demand change.

No doubt, many readers are of the opinion that I am overstating the case. So for the benefit of such simple, optimistic, misguided souls, may I refer you to an American film maker, Michael Moore. Mr. Moore is a most patriotic American. He has produced a number of documentaries, of which I consider, Capitalism: A Love Story, to be his masterpiece.

It is important to note that Mr. Moore makes no claim to be a Marxist.

In his film, he refers to the monopoly capitalists, the billionaires, the bourgeoisie, as the "one percent".

That is an expression the common people came up with, at the time of the Occupy Movement. They also referred to themselves as the "ninety-nine percent". It was a step towards class consciousness, but only a step. They were instinctively "drawing a line" between themselves, the working class, and the "super rich", the bourgeoisie.

Having said that, I have attempted to summarize the facts presented on the documentary. Bear in mind that Moore makes no reference to classes.

In his film, Moore said that he managed to get his hands on a secret Citibank memo, concerning the plan of the "one percent" to rule the world. He maintains that Citibank wrote no less than three memos to their wealthiest investors. I have done my best to summarize that which was stated in the documentary:

The bank has concluded that the United States is no longer a democracy, but a plutocracy—a society controlled exclusively by and for the top one percent of the population. As they point out, the top one percent of the citizens have control of as much wealth as the bottom ninety-five percent combined, and the gap is growing. Their biggest concern is that the ninety-nine percent may demand a more equitable share of the wealth.

The one percent is now the new aristocracy, but their big concern is that the ninety-nine percent may revolt. They lament that the ninety-nine percent still have the right to vote.

That was the conclusions that Moore was able to draw, from those memos.

Moore also makes the point that the Constitution makes no reference to capitalism, or to the right of the capitalists to make a profit.

He could have further made the point that the Founding Fathers of the United States, those who wrote the Declaration of Independence, went much further. As they stated it, "We hold these truths to be self-evident, that all men are created equal, that they are endowed with their Creator with certain inalienable rights . . . whenever any form of government becomes destructive of those ends, it is the *right of the people to alter or abolish it,* and to institute a new government." (my italics)

The Declaration of Independence gives the American people the right to *abolish any government* which does not represent them! This is a point which must be stressed to all Americans! They are one of the few people, if not the only people, who have that right! But then they have a revolutionary history, of which they can be most proud! Now it is a matter of building upon that revolutionary history.

No doubt the bourgeoisie would love nothing better than to scrap the Constitution and the Declaration of Independence. Those revolutionary documents stand for democracy, in the form of majority rule, and the capitalists are dead set opposed to such democracy.

As for those of us who have lived all our lives under capitalism, which is almost everyone, it is only natural to consider this as a normal state of affairs. Yet it is anything but the normal state of affairs. In fact, it is supremely abnormal.

Capitalism first came into existence, roughly three hundred years ago. It was the industrial revolution that gave birth to capitalism. Before that time, there were no capitalists, bourgeoisie, just as there were no workers, proletarians. The point being that capitalism is a relatively new creation. Further, it was created by people. Capitalism is not an "Act of God"! It was created by people, and it will be destroyed by people! Capitalism will be destroyed by one of the classes of people it created, the working class, the proletariat.

Marx and Engels documented the dual nature of capitalism in their landmark work, The Communist Manifesto. They reveal to all the progressive aspects of capitalism, at least in its early stages, as well as its reactionary features. This is referred to as the "dual nature" of capitalism. It is hoped that all readers will read that essential work of scientific socialism, the only true socialism.

The Communist Manifesto was written in 1848. Of particular interest is the introduction to that work, by Engels, in 1883, shortly after the death of Marx. As Engles stated:

"The basic thought running through the Manifesto—that economic production, and the structure of society at every historical epoch necessarily arising therefrom, constitute the foundation for the political and intellectual history of that epochs; that consequently (ever since the dissolution of the primeval communal ownership of the land) all history has been a history of class struggles, of struggles between the exploited and the exploiting, between dominated and dominating classes at various stages of social evolution; that this struggle, however, has now reached a stage where the exploited and oppressed class (the proletariat) can no longer emancipate itself from the class which exploits and oppresses it (the bourgeoisie) without at the same time forever freeing the whole of society from exploitation, oppression, class struggles—this basic thought belongs solely and exclusively to Marx."

Shortly after that, capitalism reached the stage of monopoly, of imperialism, which is complete reaction. This is to say that we can expect the decay of our civilization.

To counter act that, we must prepare the working class, the proletariat, for revolution and the subsequent Dictatorship of the Proletariat. With that in mind, we should stress that our "economic production" is now socialized, so that our political structure must fall into line.

Scientific socialism must be established. This is to say that it must be Marxist, as it has been proven that utopian socialism does not work. The monopoly capitalists, the bourgeoisie, the billionaires, are not about to surrender their hard stolen wealth and power, without a fight. That is the reason they must be overthrown and crushed, under the Dictatorship of the Proletariat.

I consider the quest for these huge species to be part of the revolutionary movement. We are entitled to our wildlife. It is part of our heritage.

It is frequently stated that any great scientific breakthrough, will have to be made by youngsters. Those who hold that belief, usually use the example of Newton. They point to the fact that he did all of his scientific work before the age of twenty-one. That is true, but it does not mean that the rest of us should retire our brains at an early age! I am certainly not a youngster, and am physically not capable of doing that which I did many years ago. But then my brain has not atrophied. I just hope that others, young and not so young, will be motivated to take part in a scientific breakthrough. It is not something you will ever regret.

I can only stress that this is part of the revolutionary movement. Working people must become revolutionary. This is a fine place to start.